A Treatise of

John Southall

Alpha Editions

This edition published in 2024

ISBN : 9789361475535

Design and Setting By
Alpha Editions
www.alphaedis.com
Email - info@alphaedis.com

As per information held with us this book is in Public Domain.
This book is a reproduction of an important historical work. Alpha Editions uses the best technology to reproduce historical work in the same manner it was first published to preserve its original nature. Any marks or number seen are left intentionally to preserve its true form.

Contents

THE PREFACE. ....- 1 -
A TREATISE OF BUGGS. ....- 2 -

THE PREFACE.

BEING diffident of my own Performance, and desirous it should stand or fall by the Opinion of the Best of Judges, was the Motive that induced me to make my[x] Application to that very Learned, truly Judicious and commendably Curious Person to whom it's dedicated: At the same time determining, that if he approv'd of it, I would publish it; and if he disapprov'd, that I would burn it. But it happily meeting his Approbation, it now makes its Appearance in Print: Tho' I must in Justice to him acknowledge, it could not have so done so soon, nor with such Embellishments, had he not only forwarded the Impression, but directed and order'd the Copper-plate. As it has not only his Approbation, but also, by his introducing it, the unanimous Concurrent Approbation of (those great Encouragers of things[xi] useful) the Royal Society; I hope it will not fail of meeting a kind Reception from the Inhabitants in and about this Metropolis; by whom, as such a Treatise, &c. was most wanted, for their Benefit and Ease it was at first chiefly design'd.

Tho' with such Helps as it now has, I am not without Hopes that it may extend its Qualities to distant infected Places.

I should think it a Duty Incumbent on me, and would wait on the Venerable Members of the aforesaid Society, present when my Manuscript was read, personally to return them[xii] my Thanks for the Honours conferr'd on me. But as the Names and Places of Abode of most are unknown to me, I humbly beg they will accept this Acknowledgement of them, by their

Obliged and Obedient Servant,
JOHN SOUTHALL.

A TREATISE OF BUGGS.

AS Buggs have been known to be in *England* above sixty Years, and every Season increasing so upon us, as to become terrible to almost every Inhabitant in and about this Metropolis, it were greatly to be wished that some more learned Person than my self, studious for the Good of Human Kind, and the Improvement of natural Knowledge, would have oblig'd the Town with some Treatise, Discourse or Lecture on that nauseous venomous Insect.

But as none such have attempted it, and I have ever since my return from *America* made their destruction my Profession, and was at first much baffled in my Attempts for want, (as I then believ'd, and have since found) of truly knowing the Nature of those intolerable Vermin: I determined by all means possible to try if I could discover and find out as much of their Nature, Feeding and Breeding, as might be conducive to my being better able to destroy them.

And tho' in attempting it I must own I had a View at private Gain, as well as the publick Good; yet I hope my Design will appear laudable, and the Event answer both Ends.

The late Learned and truly Valuable Dr. *Woodward*, to whom I first communicated my Intent, not only approv'd the Design, but also the Methods which, I told him, I design'd to pursue, to attain the desired Effects: and at the same time was so good to give me some useful Hints and Instructions, the better to accomplish an Affair, which he said 'twas his Opinion would be a general Good.

Not to make this Acknowledgement of his kind Assistance, would be Ingratitude to my dear deceas'd Friend.

As I had his Approbation at the beginning, had he but liv'd till now, I doubt not but the Discoveries I have made would have appear'd so considerable and useful, as might have entitled me to his farther Friendship and Assistance, in methodizing this Treatise for Publication.

But depriv'd of him, my first and greatest Encourager, I have ventur'd to let it appear in the best Dress my Capacity will admit. Should the Stile and my Manner of handling the Subject to be treated of, appear uncouth and displease, I hope the Usefulness of it to the Publick will make some amends for that Defect.

In treating on these Insects, some part of the Discourse may perhaps at first View appear surprizing, if not incredible to the Readers: But by giving them an account how I attain'd my Knowledge, and by often reiterated Experiments prov'd them to be certain Facts, they will soon alter their Opinion; and the whole, I hope, will not only be acceptable, diverting and instructive to the Readers, but also of universal Benefit to the Inhabitants in and about *London* and *Westminster*.

This Treatise being on a Subject as much wanted as any whatever, and the Pains and Trouble I have taken to arrive at my Knowledge herein, having been uncommon; it may be expected by the Curious, that I should give some of the Reasons that first induced me to undertake a Discovery so very difficult to appearance.

It may not therefore be unnecessary to acquaint such, that in the Year 1726, my Affairs requiring my going to the *West-Indies*, I had not been long there arrived, before, (the Climate not agreeing with my Constitution) I fell sick, had a Complication of the Country Distempers, lost the Use of my Limbs, and was given over by the best Physicians at *Kingstown* in *Jamaica*.

But, contrary to their Expectation, recovering a little, they advis'd me to stay no longer in a Country, so prejudicial and dangerous to me, than till I could get Shipping for *England*; and in the mean time desired that as often as I was able, I would ride out for the Benefit of the Air; which as soon as I had Strength enough, I did.

In one of my Journeys meeting with an uncommon Negro, the Hair or (rather) Wooll on his Head, Beard, and Breast being as white as Snow, I stopt my Horse to look on him; and he coming, as their way is, to beg a little Tobacco, I gave it, and enquir'd if he had been always so white-hair'd. He answer'd, no; but Age had made him so.

Observing that he moved briskly, had no Wrinkles, and all his Teeth, I told him I could not believe him to be very old, at the same time desiring to know his Age. His Answer was, he knew not himself; but this he knew, that he was one of the first Slaves brought into that Island, after the *English* had taken it in *Oliver Cromwell*'s time, and was then a great Boy.

By this account I judged, and might reasonably suppose him upwards of ninety Years of Age.

Whilst we were in discourse, he perceiving me often rub and scratch, where my Face and Eyes were much swelled with Bugg-Bites, asked if Chintses, (so Buggs are by Negroes and some others there called) had bit me? On my answering, yes; he said, he wonder'd white Men should let them bite; they should do something to kill them, as he did.

This unexpected Expression excited in me a Curiosity to have farther Discourse with him; and on my telling him, that for my part I should be extreme glad to know how to destroy those Disturbers of my Rest, and that if he would tell me how, I would give him some more Tobacco and a Bit, (a Piece of *Spanish* Money, there current at Seven-pence Half-Penny:) On this he agreed to give me a Calibash full of Liquor, which he said would certainly do it, following the Directions he gave me.

Possess'd of this, well pleas'd I went home, and tho' much fatigued, I could not forbear using some of it before I went to sleep; and to my surprize, the instant I applied it, vast Numbers did, (as he had told me they would) come out of their Holes, and die before my face.

These I swept up, threw away, and went to Bed, and had much better Rest than usual, not being Bugg-bit then, as I always was before.

But what added to my Satisfaction, and further surpriz'd me, was, that when I got up I found many more had come out in the Night and were dead.

On this, I conceiv'd so great an Opinion of the Goodness and Usefulness of this Liquor, that I was resolved to endeavour, and if possible to prevail on him to teach me how to make it; well knowing so valuable a Secret was much wanted, and would be highly useful, if I lived to return to *England*.

In order to obtain it, I got some *English* Beef, Pork, Biscuit and Beer, and some Tobacco, believing those sooner than Money or any other thing would procure from the Negro, what I so much wanted and desired.

The next day early I went, and finding my Negro in his Hutt, I asked if he could dress me some Victuals. On his replying, yes, if he had it; I open'd my Store-Bags, took out one Piece of Beef, some Biscuits and a Bottle of Beer, taking care at the same time he should see the rest. We eat a Biscuit, drank some Beer, and to dressing the Beef with some Yams out of his Plantation, he eagerly went: all Negroes being greedy of Flesh, when they can come at it; some of them not eating any for many Years together, but live altogether on the Produce of the Earth.

Whilst he was intent on Cooking and in a good Humour, I took the Opportunity of telling him, I had used and so well approved of his Liquor, that if he would learn me how to make it, all the Victuals and Tobacco in the Bags, and what Liquor we did not drink whilst I staid, I would give to him, as a Recompence for the Favour.

At first he refused, believing me (as I found by his discourse) to be a *Creolian*, whom the Negroes in general hate; but upon my convincing him I was an *Englishman*, and returning home, the good Chear prevail'd.

After eating together, into the Woods and Savannahs we went, to gather such of the Materials as grew not on his Plantation, or that he had not by him; and returning to his Hutt, to making the Liquor he went.

I remark'd well, and set down the Names, Quantities, and his way of making and mixing the Composition; which being done, all the Bottles we emptied of Beer were fill'd with the Liquor; with which I return'd to *Kingstown*, being as well pleas'd with my Discovery, as the Negro was with my Presents.

Having thus obtain'd my most material Point, I could not yet forbear going every time I rode out, to see and discourse my Negro, and never went empty-handed, being desirous to try if I could discover any thing further from him or of him, and how he attain'd the great Knowledge I found he had in the medicinal Virtues of Roots, Plants, *&c.*

He inform'd me, that during about fifty Years that he was a Slave (in which time six of his Masters had died) he oft wish'd for Death, and sought no means to preserve Life, and was then so infirm, as to be thought by his seventh Master to be past labour; and having been a good and faithful Slave, his said Master gave him his Freedom, and the piece of Ground I found him upon, to live on.

That Liberty having render'd Life more agreeable to him, he then studied all means to preserve it, and having some knowledge of things proper to preserve as well as support him, he had ever since planted, *&c.* in his Plantation, things proper for Physick as well as Food.

And indeed his Ground might be called a Physick-Garden, rather than a Provision-Plantation; for of the latter he only raised enough to support himself, of the former to supply others as well as himself, and frequently made Medicines for his sick Acquaintance and others with success. This Account I had of him from many, as well as from himself; which made me entertain so good an opinion of his Skill and Fidelity, that I ventur'd to take a Medicine made by him, by the use of which I found great Benefit in the restoring me to the Use of my Limbs.

'Twas owing to his Skill that he had thus preserv'd himself to so great an Age; and 'tis my Opinion, he had attain'd to a greater knowledge of the physical Use of the Vegetables of that Country, than any illiterate Person ever had done before him.

Believing some of the Materials not to be had in *Europe*, I procured of him a quantity, and soon after returned to *England*.

On my arrival at *London* in *August* 1727, I made some Liquor to compare with his, (which I found exactly the same) whereupon I set about destroying of Buggs, and found to my Satisfaction, that wherever I apply'd it, it brought out and kill'd 'em all. At length I advertis'd, had great business, and pleased every body, then apprehending no return of the Vermin. But yet, to my surprize, tho' I had kill'd all the old ones, young ones sometimes, in some places, would appear.

To my Liquor's being then so strong and oleous, that I durst not venture to liquor the Furniture for fear of damaging it, I at first attributed the coming of those young Buggs.

Whereupon I studied to find an Allay for that Quality; which at length, after many Experiments, and with much difficulty, I found out, and to such a perfection, that I can and do with safety liquor the richest of Furniture, as well as the Wood-work of Beds; and tho' the staining Quality be taken off, the valuable attractive and destructive parts of the Composition still retain their full Virtue.

Having gain'd this great Point, I then went on briskly, and destroyed Buggs and Nits in all Beds and Furniture wherever I came: But yet Young Ones from behind Wainscot and out of Walls would sometimes afterwards appear, get to the Beds for better quarters and food, and become troublesome.

This much perplex'd me; but these unforeseen and then unknown Difficulties, which might have discouraged a less enterprizing Genius, prompted me the more to find the Cause and Means to overcome them; which I conjectur'd might best be done, by finding their Nature and Method of Breeding, Feeding, *&c.*

In order to it, I enquir'd of many Booksellers, if any Book concerning them had ever been published; but finding none, I then made it my business to discourse as many learned, curious, and antient Men as I possibly could, concerning them; but all that ever I could gather from either, was the following Account, how and when they were first known to be brought and to breed in *England*.

Viz. "That soon after the Fire of *London*, in some of the new-built Houses they were observ'd to appear, and were never noted to have been seen in the old, tho' they were then so few, as to be little taken notice of; yet as they were only seen in Firr-Timber, 'twas conjectur'd they were then first brought to *England* in them; of which most of the new Houses were partly built, instead of the good Oak destroy'd in the old."

In the above Account of their first coming, Esquire *Pitfield* and Mr. *White*, a Chymist, Men of great Probity and Curiosity, agreed.

And as the Sap of Deal is one of their beloved Foods, this probably might be the first way they were brought. How they are still brought, I shall speak to more fully hereafter, in my Instructions to avoid them.

Finding no satisfactory Account of their Nature, Breeding, and Feeding, to be come at from others, I was resolved assiduously to set about and try all possible ways to attain it myself. My first Step was to purchase and try Microscopes, and all such Helps as could be got, and to devise such others as might contribute thereto; by which I am enabled to give you the following Descriptions of Buggs, *&c.* which the better to illustrate, is annexed from a Copper-Plate, curiously engraven by the famous Mr. *Vandergucht*, the different Species and Sizes of Buggs, as well as one correctly and finely magnified.

I was not so fond of my own Performance, as to think my Treatise merited so great an Ornament. But as the Learned and Judicious Sir HANS SLOANE had done me the Honour to peruse and approve of my Manuscript, and thought it worthy thereof, and also desired and directed the doing the said Plate by so good a Hand; I should have been wanting to myself, had I not, in deference and respect to his Opinion and Request, annex'd the same.

A Bugg's Body is shaped and shelled, and the Shell as transparent and finely striped as the most beautiful amphibious Turtle; has six Legs most exactly shaped, jointed and bristled as the Legs of a Crab. Its Neck and Head much resembles a Toad's. On its Head are three Horns picqued and bristled; and at the end of their Nose they have a Sting sharper and much smaller than a Bee's. The Use of their Horns is in Fight to assail their Enemies, or defend themselves. With the Sting they penetrate and wound our Skins, and then (tho' the Wound is so small as to be almost imperceptible) they thence by Suction extract their most delicious Food, our Blood. This Sucking the Wound so given, is what we improperly call biting us.

By often nightly watching and daily observing them with the best of Helps, having discover'd Males from Females, I determin'd, and then did put up a Pair in a Glass, as believing that to keep them the Year round, would be the only and best way to find the Nature of their breeding, feeding, *&c.* and be a means to discover what had occasion'd the Difficulties I had met with in my Endeavours and Practice of destroying them.

As the Thought was *à propos*, and the Event having answer'd Expectation, I shall now inform you of my Observations and Discoveries thereby made.

As I put up the Pair aforesaid, so did I another Pair that day Fortnight, and so every Fortnight for eighteen Months, did I put up others, with various Foods.

The first, second, third, and fourth Pair lived, but did not presently breed, it not being then their Season of so doing: But in about ten Days after I put up the fifth Pair, they all spawn'd much about the time of each other; and in about three Weeks the Spawn came to life.

Of the Spawn and different Gradations of Buggs, I shall now give you an exact Account.

The Eggs or Nits are white, and having when spawn'd a clammy glutinous Substance, they flick to any thing spawn'd upon, and by Heat come to Maturity and Life. The Eggs are oval, and as small as the smallest Maw-feed.

Buggs of one day old, are white as Milk.

At three days old, are Cream-colour'd.

At one Week old, are Straw-colour'd.

At two Weeks, are of the same Colour, with a red List down the Back.

At three Weeks, List Copper Colour.

At four Weeks, List Browner.

At five Weeks, List deeper Brown.

At six Weeks, the Sides brown, and the List hardly discernible.

At seven Weeks, they come to be of their proper Colour, all over brown.

At eight Weeks, they grow bigger.

Nine Weeks, Ditto.

Ten Weeks, Ditto.

At eleven Weeks, they are full grown.

Vide the Plate done from *Europeans* bred: under which is a single one longer and larger, than our full-grown, being a full-grown *American* bred. 'Tis needless to give the Gradations of that Species, because when they spawn and breed here, the Young degenerate, and are of the *European* Size.

As I wrote down the Time I put up all Pairs for breeding, and also the Times they spawn'd, and observ'd and set down the Numbers they generally spawn'd; I found by my account of above forty Pair so put up with various Foods, not only their best-beloved Foods, but also their Method of Breeding; of which, to render my Observations of publick Service, I shall give you an account.

Viz. Their beloved Foods are Blood, dry'd Paste, Size, Deal, Beach, Osier, and some other Woods, the Sap of which they suck; and on any one of these will they live the Year round.

Oak, Walnut, Cedar and Mahogoney they will not feed upon; all Pairs I put up with those Woods for Food, having been soon starved to death.

Wild Buggs are watchful and cunning, and tho' timorous of us, yet in fight one with another, are very fierce; I having often seen some (that I brought up from a day old, always inur'd to Light and Company) fight as eagerly as Dogs or Cocks, and sometimes one or both have died on the Spot. From those so brought up tame, I made the greatest Discoveries.

They are hot in Nature, generate often, and shoot their Spawn all at once, and then leave it, as Fish do.

They generally spawn about fifty at a time, of which Spawn about forty odd in about three Weeks time usually, (but sometimes two or three days more or less, according as the Weather proves more or less hot) come to life; the Residue proving addle, as do often the Eggs of Hens, *&c.*

Thus they spawn four times in a Season; *viz.* in *March*, *May*, *July*, and *September*: by which 'tis apparent to a Demonstration, that from every Pair that lives out the Season, about two hundred Eggs or Nits are produc'd; and that out of them, one hundred and sixty, or one hundred and seventy, come to Life and Perfection.

Some of the first Breed I have known to spawn the same Season they were hatched; but so few in Quantity, and those so weakly, that the Winter killed them.

I have also observed that in Rooms where constant Fires have been kept Night and Day, they have been so brisk and stout as to spawn in the Depth of Winter: but of all the Spawn I ever saw between *September* and *March*, not one ever came to Life.

This plainly evinces, that Natural Heat only produces Life in the Spawn, and that Artificial cannot.

Thus having shewn plainly the Number each Pair annually produce, I hope their great Increase is so sufficiently accounted for, that it need no more be wonder'd at.

And having also shewn their seven Months Season of Breeding, if 'tis admitted, as I think 'tis plainly apparent, that in the other five Months, *viz.* from *September* to *March*, when there is no such thing as Spawn but what is addle, and consequently cannot come to Maturity; it then naturally follows, that the Winter is the best Season for their total Destruction: which I shall make more fully appear presently, but must first refute two vulgar Errors.

The first is, That many People imagine they are dead in Winter. This is a Notion so absurd, that it would hardly be worth mentioning, had I not by Experience found it had prevail'd with many People of Sense and Learning, as well as the Vulgar and Illiterate. The many Experiments by me shown at the Hospitals in the hardest Frosts last Winter, and in the Houses of the Nobility and Gentry, and to Sir HANS SLOANE the 30th of *December* 1729, will, I hope, be deem'd a sufficient Refutation of that Error: For in the coldest Seasons the Application of my Liquor with a Feather only, made the Vermin bolt out of their Holes, and die before their faces.

This they will do all the Year round in the coldest or hottest Weather. And I have seen, and do assert, they do bite in the cold as well as hot Seasons: but as our Blood is not so apt to inflame in Winter as in Summer, their Bites make but little Impression, and are consequently the less regarded.

The second and most prevailing Error is, That Buggs bite some Persons, and not others: When in Reality they bite every Human Body that comes in their way; and this I will undertake plainly to demonstrate by Reason.

It is generally observ'd and granted, that a Person under an ill Habit of Body, if he receives a small Cut or Wound, so slight as to be at first thought a Trifle, such Person's Wound by reason of such ill Habit shall be attended with Inflammations and other dangerous Symptons, and be longer under Cure than Wounds, which when first receiv'd were larger, and consequently thought more dangerous. These Wounds shall be immediately healed on Persons in good Habit of Body, such good Habit preventing any Inflammations.

And as Fevers, and Swellings attending and prolonging the Cure of Fractures, are accounted for the same way; why may it not by the same parity of Reason be admitted, that the Bite or Wound of a Bugg should swell and inflame such only whose Blood is out of order; and tho' they do bite, cause no Inflammations on any in right order of Blood?

The best Reason which can be given in support of this Error, is, That where two Persons lie in one Bed, one shall be apparently bit, the other not.

Buggs indeed, where there are two Sorts, may feed most on that Blood which best pleases their Palate; but that they do taste the other also, to me is apparent: And whenever that Bedfellow who is most liked by Buggs shall lie from home, the other will so sensibly feel the effects to be as above, that they will no longer think themselves bite-free.

Of this I am sensible, that I daily am bit when practising and at work in my Business, destroying them; and as they never swell me but when out of order, from thence I infer, that not only myself, but all such who are among Buggs, and do not swell with their Bites, are certainly in good Habit of Body. But to return to my Subject.

Having shewn that they not only live in Winter, but asserted that to be the best Season for their total Destruction, I must further observe, that few People caring to trouble themselves about Buggs but when they themselves are troubled by them, having confin'd the Attempts for their Destruction chiefly to the Breeding-Season, has been the sole Reason why the best Efforts for their Destruction have fail'd.

I do admit innumerable Quantities have been destroy'd, and much good has and may be done in Summer: but should every old Bugg then be destroy'd, you are yet not safe; for the Nits behind Wainscot and in Walls, which cannot be come at, will by heat come to life, and your work is partly to be done over again.

Whereas, on the contrary, if cleared out of Spawning-time, there is a certainty, as there is then no Nits, that their Offspring cannot plague you thereafter.

'Tis for this Reason I warrant what I do in Winter; which I cannot pretend to do in Summer.

In Summer indeed I do destroy all Buggs and Nits too in Beds and their Furniture, but Buggs only behind Wainscot and in Walls: for tho' my Liquor has an attractive as well as the destructive Quality, and thereby does bring out and destroy every live Bugg; yet their Nits being inanimate, cannot come to the Liquor, nor the Liquor at them. Such Nits therefore will come to life by heat, and quit the Walls and Wainscot for better Quarters and Food in the Bed, and so become troublesome to you.

Having thus given, I hope, a satisfactory Account of this nauseous, venomous Vermin, I shall proceed to shew how they are daily brought to *England*, and into Houses commonly; then give some necessary Cautions how to avoid them, and Directions how to destroy them.

As these Insects abound in all foreign Parts, especially in hotter Climates more than they do here; 'tis on that account all Trading Ships are so over-run with them, that hardly any one thing, if examin'd, will be found free.

And as by Shipping they were doubtless first brought to *England*, so are they now daily brought. This to me is apparent, because not one Sea-Port in *England* is free; whereas in Inland-Towns, Buggs are hardly known.

This Metropolis therefore, as having the greatest Number of Shipping, has had the greatest Number imported, and consequently bred in it.

You that are free, and would avoid a foreign Supply in your Houses, examine well all things from on board Ships before you admit them into Lodging-Rooms. Chests and Casks, Linnens, and Paper, being stiffened with Paste, afford them Food, and are consequently most dangerous.

If you have occasion to change Servants, let their Boxes, Trunks, &c. be well examin'd before carried into your Rooms, lest their coming from infected Houses should prove dangerous to yours.

Examine well all Furniture that comes in, before you set it up, Beds especially; which I recommend should be plain, and as free from Wood-work as possible, and made to draw out, that the Wainscot and Walls may be better come at, to clear them from Buggs and Dirt.

Deal Head-Boards, and Head-Cloths lined with Deal, or Rails of that Wood, avoid.

Also Beach-Bedsteds, for all such afford them much Harbour and Food.

If for Ornament you use Lace, let it be sewed, not pasted on, for Paste they love much.

Oak-Bedsteds, and plain Wainscot Head-Boards, and Tester-Rails of that Wood, allow them the least Harbour, and no Food; such therefore I recommend.

If you put out your Linnen to wash, let no Washer-woman's Basket be brought into your Houses; for they often prove as dangerous to those that have no Buggs, as Cradles, and Bugg-Traps made of the same Wood, often do to those that have them: for the Wood they are made of, affords them much Convenience of Harbour, and great Nourishment.

Upholsterers are often blamed in Bugg-Affairs; the only Fault I can lay to their Charge, is their Folly, or rather Inadvertency, in suffering old Furniture, when they have taken it down, because it was buggy, to be brought into their Shops or Houses, among new and free Furniture, to infect them.

Builders are more blameable than they: for in Houses built for Sale, old Wainscot-Doors, Chimney-Pieces, &c. are oft put up for Cheapness, painted over, and pass for new; thus the Houses in *Hanover* and *Grosvenor-Squares, &c.* were supplied before inhabited.

In taking of Houses, new or old, and in buying Bedsteds, Furniture, *&c.* examine carefully if you can find Bugg-marks. If you find such, though you see not the Vermin, you may assure yourself they are nevertheless infected.

To such as have, and would destroy them, I shall now proceed to give full Directions. In order to do it effectually, and to lessen your trouble, the first necessary thing to be known, is their Marks.

Buggs, tho' nasty to us, are so cleanly to themselves, that they quit their places of Harbour to come out and dung, and their Excrements leave a Mark or Stain like that of a Fly, but somewhat blacker; and wherever you see such Marks, if on Wood, look for the nearest Crevise, Knot, or Streak; if on Walls, for the nearest Crack or Hole; if on Furniture, for the nearest Seam, Lace, Tape, or Fold, and there assuredly are the Vermin, and there apply the Remedy.

In Winter-time, few, if any, are to be found by day-light in the Furniture of a Bed; but in the Wood-work, Wainscot, or Walls only.

In the Summer they are all over, and every Lace, Tape, Seam and Fold must be examin'd, as well as the Crevises, Joints, and Carving in the Wood-work, for the Marks, and the Remedy applied accordingly.

In Winter-time, though they lie in pretty close Quarters, yet are they easily destroy'd with any thing that will attract or entice them to it.

If no such thing you have, give me leave to recommend my Liquor; on the Application of which, at all Seasons of the Year, they will come out, and immediately die before your Face.

In Summer they lie in more open Quarters, and spread, and then are not in any measure to be reduced, but by such Liquors as you may safely touch the Furniture with all over: if none such you have, you may depend that mine will not stain or any way hurt the richest Velvet, Silk, or Stuff, not even Scarlet, which almost every thing else will do.

On account of these excellent Qualifications, the Liquor has its Name of *Nonpareil*, and of this, if minded to do it yourself, you may have a Bottle for 2 *s.* sufficient for a common Bed, with plain Directions how to use it effectually.

If the Trouble of doing it your selves be disagreeable to you, you may have it expeditiously done by me or my Servants, and your Beds, or such Part as is necessary, taken down and put up again in full as good, if not better Order, than they were before, and alter'd, (if I see Opportunity or Occasion) and made to draw out, on my usual easy Terms.

As I have occasionally mention'd what Sort of Beds I would have you avoid, give me leave to add and assure you, that Beds may be made full as warm as usual, and very ornamental, and yet be so very handy, as that any one of your own Servants might take all down and clear them of Buggs, Dirt, and Dust, and put them up again in a quarter of an Hour; and to this Fashion most Beds may be alter'd, several Persons having lately done so by my Directions to their very great Satisfaction.

Those that have a mind to have new Beds thus made, or old ones alter'd, are to observe, That the Head-posts of the Bedsted are to be no higher than just to support a Wainscot Head-board, the Tester-lath supports the Rod as usual; in the Rail are to be nine Holes drilled in, but not quite thro'; the two at the Head, to take off and hang on, (at Pleasure,) two Upholders drove into the Wainscot or Wall; in the other seven, thro' Eye-let Holes, at proper Distances in the Tester-cloth, are to be seven Balls or carved Branches to keep the Tester-Cloth tight; to which the Head-cloth, and inside and outside Vallens are to be fixed: so that by taking the Lath of the Upholders, and taking out the Balls, they all come off together; and thus made, may be commodiously and immediately clear'd, clean'd, and put up again, to fasten on the Head-board: And keep your Head-cloth down tight in its Place and Form. You have Hooks and Eyes to take on and off at Pleasure.

Persons wanting to be clear'd and kept free of those nauseous venomous Vermin, shall be attended by the Author on the following Terms, *viz.*

To clear a Bed-sted with Moulding-Tester, Wood Head-Cloth, Head-board and its Furniture, 10 *s.* 6 *d.*

Bed-steds with single-rais'd Tester, Moulding, Head-Cloth, Board and Furniture, or Chair-beds and Furniture, 8 *s.* each.

Bed-steds with ditto Tester, plain Head-cloth, cover'd Head-board and Furniture; and Field-beds and Furniture, at 7 *s.* each.

Four-post Bed-steds, or Canopy-beds, with plain Furniture, 6 *s.* each.

Press-beds, Chest of Drawers Beds and Bed-steds, without Furniture, 5 *s.* each.

Wainscot Walls, Hangings, *&c.* behind and near the Bed-sted are clear'd with the Beds at the above Prices: but if spread all over the Room and Furniture, then an additional Price is expected.

For Expedition and Safety, and to prevent Trouble to his Customers, or Impositions on them or himself, the Author takes his own Servants with him, to take down and put up such Parts of Beds, Wainscot, Hangings, *&c.* as he finds necessary; and always puts them up in full as good, if not better Order, than he finds them. Of his Servants he has good Security, and will be answerable to his Customers, for their Fidelity.

N. B. If he any ways damages the Furniture, he will pay for the same.

Persons about taking Houses, Lodgings, or buying Furniture, paying for Surveying, shall be attended, and at first View be justly and truly inform'd if the Premisses be Buggy, or free from Buggs, by

JOHN SOUTHALL,

At the *Green-posts* in the *Green-walk* near *Faulcon-stairs, Southwark.*

FINIS.